Fact File
DEADLY & DANGEROUS

Author & Principal photographer: Michael Cermak

Steve Parish™

KIDS

Introduction

WHAT IS A DANGEROUS ANIMAL?

Different animals have different ways of defending themselves. Some use their teeth, tusks, claws, horns or antlers; others use venom, *poison*, acids and other chemicals to protect themselves or to kill their prey. If given a choice, almost all animals will try to escape humans rather than attack. However, any animal that is able to kill, injure or even cause discomfort to a person is regarded as dangerous. In this book we have placed dangerous animals into four different groups — **deadly**, **dangerous**, those that are **harmful** or can cause injury or an unpleasant reaction, and those that are **disease carrying**.

The **deadly** group is made up of animals that have caused death by biting, stinging, stabbing, scratching or poisoning people, either during defence, in a deliberate attack, or when eaten. Some of these animals may surprise you, although in some cases the death may have occurred due to a freak accident or allergy, rather than an attempt to kill. The **dangerous** group is made up of animals that may injure (or, in very rare cases, even kill) people but usually do so only if threatened. The **harmful** group is made up of animals that cause a painful reaction if stepped on, touched or brushed, and (for animals like the magpie) those that can harm us if we are in their territory. In other words, these animals are not really doing anything except defending themselves from us. The **disease carrying** group is made up of creatures that can give people diseases, or that are parasites. Mosquitoes, some flies, flying-foxes, rats, feral pigs and some fishes fall into this group.

THE DIFFERENCE BETWEEN DEADLY AND DANGEROUS

Apart from some sharks and estuarine crocodiles, animals do not normally attack people unless provoked. More often than not, it is

The Portuguese man-o-war (or bluebottle) has an unpleasant sting.

Sydney funnel-web spiders have long, large fangs.

Magpies can be aggressive when nesting.

The giant bulldog ant's mandibles can give a nip, but the sting on its tail is very dangerous and has killed people in Australia.

our behaviour that causes problems with animals — sometimes we may not even notice that we have disturbed an animal in its environment. It is important to understand that just because a certain animal is deadly and can harm or kill someone, that doesn't always mean that the animal is very dangerous. For example, the western taipan, *Oxyuranus microlepidotus*, the most *venomous* snake in the world, lives in mostly unpopulated areas of the Australian outback where few people ever go. So this snake is very rarely seen and even when it is seen it is shy and moves quickly into hiding.

However, the eastern brown snake, *Pseudonaja textilis*, is found in most parts of the continent, often living close to people in farms and suburbs. It does not take much for this snake to become annoyed and aggressive. Despite having short *fangs* and venom that is less toxic than the western taipan's venom, the eastern brown snake is more dangerous to humans.

Deadly and dangerous animals are fascinating. By understanding how these animals live and behave, how they attack and the effects of attack, you should find that it is easy to stay safe from these animals.

FACTS ON THE FACT FILE RANGE

Fifty-six *species* are featured in this fact file. The introduction to each species helps you discover fascinating facts about a dangerous animal. The information underneath tells you how big the animal grows, the likely effects of an attack, how to avoid an attack and where to watch out for the animal. Words throughout the book that are in italics are included in the glossary on page 47 to help you increase your knowledge.

Contents

The estuarine crocodile, also known as the saltwater crocodile, is the largest species of crocodile in the world.

Crocodile attacks have awful, even fatal, results but the chances of being attacked by a crocodile are low. There are only about 4000 estuarine crocodiles living in northern Australia, but around 300,000 tourists a year visit areas where these crocodiles live. A crocodile attack might not happen very often, but with so many humans in their *habitat*, close encounters with Australia's top *predators* do sometimes occur.

SIZE: Estuarine crocodiles grow to a huge 7 metres in length. However, the largest crocodile ever killed in Australia measured less than 6 metres.

NUMBER OF HUMAN DEATHS: Since 1985, estuarine crocodiles have killed 11 people in Australia. In Queensland there have been 15 crocodile attacks in the past 21 years — only five of these were fatal.

SEVERITY OF AN ATTACK: The size and body weight of an adult estuarine crocodile is enough to harm anyone! If the bite and the enormous pressure of the crocodile's jaws don't kill its victim, the crocodile will drag its prey into the water and drown it. Using its powerful jaws to secure its prey, the crocodile folds its legs along its body and quickly spins around and around. This "death roll" throws its prey off balance.

Under the water, a crocodile can hold its breath for much longer than any other land animal, including humans. The crocodile's teeth interlock when the jaws are shut, so the bite itself causes terrible damage. If the victim struggles, the crocodile violently shakes its head from side to side, sometimes tearing its victim apart.

AVOIDING AN ATTACK: Many estuarine crocodile habitats in northern Australia have signs warning visitors to take care. It is extremely important to take these warnings seriously, even if the creek or waterhole seems too small or too shallow for a crocodile to live there. If you are fishing or camping in crocodile country, stay away from the water's edge, especially at night, which is when crocodiles are most active.

An estuarine crocodile's pointy teeth grow back if they fall out.

Signs warn visitors to be careful in areas where crocodiles are found.

Crocodiles are experts at stealth, floating near the surface like logs.

Never throw fish scraps or food into the water or on the bank, as this may attract hungry crocodiles. If you are camping near water for a long time and must go to the water's edge, choose a different path to a new spot on the bank each time. Most animals come to drink at the same time and at the same place every day — crocodiles notice this and move in closer each time until they are within striking distance. The same thing can happen to careless humans!

WHERE TO WATCH OUT: Estuarine crocodiles live in tropical north Australia, from the west coast of Western Australia, across to the tip of Cape York Peninsula and down to Rockhampton in Queensland.

These fearsome crocodiles are also called "saltwater crocodiles" because, unlike their close relatives the freshwater crocodiles, "salties" can live in saltwater. However, the name estuarine crocodile is more suitable because they do not only live in saltwater — they can also live in freshwater creeks, rivers and billabongs far away from the sea.

It is often extremely hard to see saltwater crocodiles because they are shy, secretive animals that are extremely well *camouflaged*. Always look for crocodile tracks on muddy banks, or "slides" where they get in and out of the water. And remember — if you can't see the bottom, don't even go near the water's edge!

Hungry crocodiles make a meal of just about anything, even bats!

If you see tracks like this in mud or sand, be extremely careful.

Western (inland) taipan *Oxyuranus microlepidotus*

The western taipan's head is usually black but may also be brown.

The western taipan, also known as the "inland taipan", "fierce snake" or "small-scaled snake", is the most venomous snake in the world! Its venom is twice as deadly as the venom of the eastern taipan. However, the western taipan lives in largely uninhabited areas of Australia, so most humans never come across this snake. In fact, this snake is so hard to find that it was only discovered in the late 1970s!

SIZE: The western taipan is a large snake that measures about 2–2.5 metres.

NUMBER OF HUMAN DEATHS: There have been no reported deaths caused by a bite from a western taipan. It is regarded as "deadly" only because of its very powerful venom.

EFFECTS OF VENOM: This snake carries the most potent venom in the world. It contains *neurotoxins* that attack the central nervous system. The victim is quickly *paralysed* and then dies if untreated. If bitten, seek medical help immediately.

AVOIDING AN ATTACK: It is actually quite easy to avoid the western taipan because it is a very shy creature and will try to escape rather than attack. However, the western taipan will attack if handled. Avoid poking about in cracks in the soil because this snake likes to hide in such places and only comes out to lie in the sun and hunt. The western taipan has adapted to survive in harsh, dry environments where there are few trees or bushes. However, after the rainy season, the Channel Country becomes covered with thick, low shrubs that make a nice shady habitat for the western taipan. If you are in the outback, make sure you stay on the road and do not walk across any overgrown paddocks.

WHERE TO WATCH OUT: The western taipan is found in central Queensland, south-east parts of the Northern Territory, western New South Wales and north-east South Australia.

A western taipan's skull, showing its long fangs and teeth.

deadly

Eastern (coastal) taipan *Oxyuranus scutellatus*

Although not as venomous as the western taipan, the eastern taipan is more aggressive.

The eastern taipan is probably Australia's most feared snake. It is aggressive when threatened, huge, has the longest fangs of all Australian snakes, injects enormous amounts of very powerful venom, moves and strikes extremely fast and is easily disturbed!

SIZE: Eastern taipans are usually about 2 metres long, but sometimes they can be more than 3 metres long!

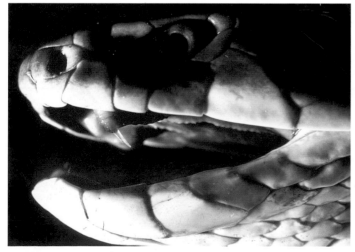

The eastern taipan's fangs can grow to 1.3 centimetres long.

NUMBER OF HUMAN DEATHS: Medical records show that at least six people died from a taipan bite before an *antivenom* was made in 1955.

EFFECTS OF VENOM: A taipan strikes accurately, sometimes several times in a row. Its long fangs (up to 1.3 centimetres long) inject venom deep into the flesh. The venom is highly neurotoxic. It causes very little local pain or swelling, but paralysis and other serious symptoms can quickly follow. Getting to the nearest hospital for fast, accurate first aid treatment is absolutely essential in saving the life of someone who has been bitten by this deadly snake.

AVOIDING AN ATTACK: Taipans are shy snakes, always trying to slither away from an approaching person. If you see an eastern taipan in the bush, stay away from it and walk off in another direction. Threatened snakes attack very quickly. When you go bushwalking, take a pressure bandage with you and wear protective clothes, such as long pants and boots.

WHERE TO WATCH OUT: Eastern taipans are found along the east coast of Queensland, from the New South Wales border all the way to the tip of Cape York Peninsula. They also live in the northern parts of the Northern Territory and Western Australia.

deadly

The eastern brown snake comes in many different shades of black, grey, yellow or brown, with an orange-spotted, cream belly.

The eastern brown snake is one of the most common snakes in eastern Australia and it is responsible for more than half of all snake bites in the whole country. Snakes rarely attack humans without a reason, but it does not take much to upset an eastern brown snake! They have been known to lunge toward passing cars, and may chase people for some metres in the bush and in suburban backyards, and attack dogs and cats.

Eastern brown snakes eat mainly mice and small rats, which is why they like farms and other places where people live. Coming across an angry eastern brown snake is a scary experience. The snake raises the front part of its body into an "S"-shaped position and hisses loudly with its mouth open while slithering towards the intruder.

When the eastern brown snake gets close enough to bite, it stops and waits to see what happens next. If the intruder moves away (or stays very very still) chances are that the brown snake will calm down and flee. Moving away from the snake is, of course, the best thing to do!

SIZE: This snake normally grows about 1.5 metres long, but can grow to a maximum length of 2 metres.

NUMBER OF HUMAN DEATHS: Since 1981, nineteen people have died from an eastern brown snake's bite.

Although the venom of the eastern brown snake is extremely toxic, only a small amount can be injected due to the snake's short fangs.

EFFECTS OF VENOM: The eastern brown snake has one of the most toxic venoms of all Australian snakes and has the second most potent venom of all land snakes in the world (only the western taipan has more powerful venom). Luckily, the eastern brown snake has very short fangs, so it often only injects a small amount of venom.

However, the eastern brown snake's venom causes very serious health problems. The venom contains a mixture of two contrasting chemicals called *procoagulants* and *anticoagulants*. Procoagulants cause a victim's blood to clot and anticoagulants have the opposite effect, making the blood thin and runny. After the victim is bitten, its blood clots. However, when the procoagulants are all used up, the anticoagulants start to work, causing internal and external bleeding. The result is general sickness, paralysis, early collapse and possibly death. Specific antivenom is available in all major hospitals in areas where eastern brown snakes live. Interestingly, the eastern brown snake has a very high rate (75%) of "dry bite" — that is, a bite that does not inject venom.

There are seven species of brown snakes in the genus *Pseudonaja*:

Dugite *Pseudonaja affinis*

Speckled brown snake *Pseudonaja guttata*

Peninsula brown snake *Pseudonaja inframacula*

Ingram's brown snake *Pseudonaja ingrami*

Ringed brown snake *Pseudonaja modesta*

Western brown snake *Pseudonaja nuchalis*

Eastern brown snake *Pseudonaja textilis*

THEY ARE ALL CONSIDERED DANGEROUS

This eastern brown snake is ready to strike.

AVOIDING AN ATTACK: Wearing long pants and solid shoes will provide good protection against the eastern brown snake's short fangs. To help warn snakes away, stamp your feet hard against the ground when bushwalking. Snakes will feel the vibrations on the surface of the ground and will slither away.

WHERE TO WATCH OUT: Eastern brown snakes are common in eastern Australia, from the southern-most point of Victoria to the tip of Cape York Peninsula. In fact, along the east coast the only places where eastern brown snakes are not found are on top of high mountains and in rainforests.

These snakes blend in well with dry grass or leaf litter and often seek cover under flat objects such as fallen trees, bark or rocks. In suburban areas, eastern brown snakes hide under old building materials or any other rubbish that may be left around the house. Keeping the backyard tidy means there is less chance of an eastern brown snake moving in!

Tiger snake *Notechis scutatus*

This tiger snake is ready to strike! Before striking, a tiger snake will pull its neck into an "S" shape.

Tiger snakes may be found in large groups, usually in river valleys where they have plenty of frogs to eat. Unfortunately, Australia's first European settlers also liked to settle near rivers, which provided the communities with a permanent water supply. Because snakes and people lived close together, snake bites were common and often ended in death. The number of deaths has decreased since 1930, when the first tiger snake antivenom was developed.

SIZE: On average, tiger snakes grow to 1.2 metres but some can grow to 2 metres long.

NUMBER OF HUMAN DEATHS: It is hard to find accurate figures, but tiger snakes are considered the second most common cause of snake bite related deaths in Australia.

EFFECTS OF VENOM: The venom of a tiger snake is the fourth most *lethal* of all Australian snake venoms. It is highly neurotoxic and quickly affects the nervous system. It also contains procoagulants, which cause the blood to clot. A person bitten by a tiger snake may suffer headache, nausea, breathing difficulties and convulsions. In serious cases, liver failure, paralysis and death may follow.

AVOIDING AN ATTACK: Tiger snakes are not aggressive by nature. They prefer to quickly escape rather than attack. When cornered or annoyed, a tiger snake flattens its body and hisses loudly, thrashing its body from side to side as a warning. It is always best to move away and leave a tiger snake alone.

WHERE TO WATCH OUT: Tiger snakes live in southern parts of Australia (including Tasmania) where they *hibernate* during the winter. They are only active during the warmer months of the year, particularly early morning and late afternoon when they lie in the sun. They also like to rest among tall patches of grass, where they are hard to see. Always wear long pants and solid shoes when bushwalking in tiger snake country, and remember to always look before sitting down on the ground.

The yellow and black stripes give the tiger snake its name.

deadly

Rough-scaled snake *Tropidechis carinatus*

Although it is rarely seen, this small snake has been responsible for many serious bites. The rough-scaled snake is very fast and aggressive — ready to bite without any warning. This snake can also be easily mistaken for the harmless keelback, *Tropidonophis mairii*, which looks similar. Both species have *keeled scales*, are about the same size, and often share the same habitat.

SIZE: On average, this medium-sized snake measures less than 1 metre long.

Above and top: The rough-scaled snake looks quite similar to the harmless keelback snake, but the rough-scaled snake is venomous.

NUMBER OF HUMAN DEATHS: At least one person has died from this snake's bite and many victims need to go to hospital for medical treatment.

EFFECTS OF VENOM: A rough-scaled snake strikes rapidly, often biting quickly several times. The bites leave very small puncture marks on the skin. This snake injects its highly neurotoxic venom into the victim's body tissues, leaving only tiny amounts of venom on the surface of the skin. What makes the rough-scaled snake extremely dangerous is that a bitten person may collapse just a few minutes after the bite. That means that if the victim is alone and far away from medical help, the bite can be fatal. If medical treatment is available, it may be possible to give the victim tiger snake antivenom because the venom of these two snakes is similar.

AVOIDING AN ATTACK: These snakes live in coastal ranges and rainforests from central New South Wales to South-East Queensland and the Wet Tropics. If you are visiting these places, wear sensible clothing for bushwalking, such as boots or thick shoes and long pants, to help protect you from rough-scaled snake strikes. Never handle or annoy this snake under any circumstances — it will fight back!

WHERE TO WATCH OUT: These snakes are found in a few locations: on the central coast of New South Wales to South-East Queensland, in tropical Queensland between Tully and the Carbine Tableland and in the Wet Tropics from Bluewater Range to Thornton Peak. Rough-scaled snakes are active both day and night, and like thick forest undergrowth where their colour helps them to blend in. They like to climb up shrubs and ferns to enjoy the morning or afternoon sun. Being climbers, they are often at knee or waist level to bushwalkers and are easily overlooked.

The really interesting thing about this snake is that it injects different amounts of venom depending on the situation. To make sure it kills its prey, a rough-scaled snake may inject more venom when hunting than when simply striking out in defence.

deadly

This common death adder is extremely well camouflaged among the bark of a grass-tree.

The common death adder is one of the most feared snakes in Australia. It is also the country's fifth most venomous snake. Common death adders are very difficult to spot in the bush. They rely on camouflage to help blend in with their surroundings and stay safe from intruders.

The common death adder's bite is not always effective — sometimes the victim ends up with a "dry bite". Studies show that "dry bites" are warning bites, which are usually unprovoked. Once the snake is provoked, a "dry bite" is very rare. In any case, any bite from a death adder must be treated as extremely serious.

SIZE: This is the largest of the four death adder species. The average size of an adult is 65 centimetres long, but the record size for this species is 1.1 metres!

NUMBER OF HUMAN DEATHS: Common death adders are responsible for at least five deaths in Australia. Before an antivenom was developed, 60% of all bites from this snake were fatal!

EFFECTS OF VENOM: A person bitten by a common death adder will feel sick (or nauseous) at first and may feel a slight burning sensation around the bite. The victim later develops blurred vision, breathing difficulties and paralysis. How quickly these symptoms develop depends on a number of things — the size and health of the snake, the size and health of the bitten person, which part of the body has been bitten, and how quickly and how properly first aid treatment is applied to the victim.

The common death adder's skull, showing its long fangs.

AVOIDING AN ATTACK: This snake looks very different from other venomous Australian snakes. It is short with a fat body and a broad, triangular-shaped head. Also, unlike any other Australian snake, the common death adder has a very special tail. Because the common death adder doesn't move very fast, it prefers to *ambush* its prey, relying on camouflage and using its tail as a lure to draw prey close. The tip has a different colour and looks like a worm with a small spike at the very end. The snake attracts prey by wriggling the tip of its tail in front of its head. Thinking this is a tasty worm, a curious mouse, bird or lizard will move in for a closer look. When its prey is within striking distance, the death adder quickly attacks and kills the animal. The death adder cannot use the spike on the end of its tail to sting with — that end of the death adder, unlike its head, is harmless!

The common death adder is active mainly at night, especially on warm spring and summer nights when males go in search of females. However, during the day, this snake is very hard to spot as it lies very still among leaf litter with just its head poking out. This snake's scales are quite dull and its colour blends in well with its surroundings.

When planning a bushwalking or camping trip, it is always wise to check what kind of snakes live in the area. In some places, this species of death adder is much more common. Common death adders are not aggressive. They bite only when threatened or surprised. The common death adder may thrash from side to side, but it will not chase people or jump. If disturbed, a common death adder flattens its body, making it look much wider than it really is. When it does this, the bright colours of its skin can be seen between its scales. The best way to avoid being bitten by a death adder is to wear long pants and boots in the bush, and to walk away if you see this snake.

WHERE TO WATCH OUT: Common death adders are found in New South Wales and Queensland (except for Cape York Peninsula), eastern Northern Territory and southern South Australia. They can be found in forests, grasslands or deserts.

The colour of the common death adder varies — they can be red, brown, grey or a mix of these colours.

The king brown's head is wider than that of other venomous snakes.

The king brown snake is a large, solid snake that is able to inject large amounts of venom. On average, a king brown snake produces 180 milligrams of venom, but big snakes can generate as much as 1350 milligrams of venom! Although its venom is not extremely powerful, the volume of venom that the king brown can inject puts this snake on the list of Australia's very dangerous snakes.

SIZE: An average sized king brown snake is about 2 metres long, but some grow even bigger than this.

NUMBER OF HUMAN DEATHS: Information about deaths caused by the king brown snake is unreliable. This snake is not easily identified, so perhaps many bites were thought to be from another species — a fatal mistake because only one antivenom is known to work well. We do know that at least one person died from a king brown's bite in Western Australia in the 1960s.

EFFECTS OF VENOM: Venom from a king brown snake is different to most other venoms. It has several toxins but it also has strong anticoagulants, chemicals that stop the victim's blood from clotting.

While most other snakes bite and let go, the king brown snake chews onto its victim, injecting a lot of venom. For this reason, any bite from a king brown snake could be lethal. Many people think that the king brown snake is a kind of giant brown snake, or that any large brown coloured snake must be a king brown. The fact is, the king brown snake is actually a member of the black snake group (genus *Pseudechis*) and is not related to the brown snakes at all! This fact is extremely important because brown snake antivenom does not neutralise king brown snake's venom. When tiger snake antivenom was used in the past to try to neutralise the king brown snake's bite, only one in every ten victims survived. There is now a black snake antivenom available to counteract king brown bites.

AVOIDING AN ATTACK: King brown snakes are usually active at night but also like to lie in the morning and afternoon sun. When surprised, they flatten their entire body, hissing and thrashing from side to side. When they are doing this, you should definitely get away from the angry snake as quickly as possible. When camping in the outback, always use a torch when you are walking around the campsite at night and never walk around with bare feet.

WHERE TO WATCH OUT: King brown snakes are found throughout Australia, except for the eastern coastal areas, the south coast, Victoria and Tasmania. These snakes like to live alone, which suits their lifestyle. They are cannibalistic, which means they eat other members of their own species, and they also feed on other snakes and lizards.

In some parts of the Northern Territory, particularly at Fogg Dam near Darwin, king brown snakes share their habitat with western brown snakes and also with water pythons. All of these snakes are a similar brown colour and can easily be mistaken for one another. If you ever find a snake that you cannot identify, the best rule is always to treat it as if it were very dangerous. As the old saying goes, "it is better to be safe than sorry".

King brown snakes live across Australia in many different habitats.

Eastern small-eyed snake *Cryptophis nigrescens*

Eastern small-eyed snakes are usually only 50 centimetres long.

SIZE: The average size of this snake is 50 centimetres, but the maximum size recorded is 1.2 metres.

NUMBER OF HUMAN DEATHS: One person died from the eastern small-eyed snake's bite in 1965.

EFFECTS OF VENOM: The victim may feel a stinging around the bite area, slight swelling and a headache, although often these symptoms do not appear until a few days later. If left untreated, an *allergic* reaction to the venom may occur, which could be more dangerous than the venom itself.

AVOIDING AN ATTACK: Eastern small-eyed snakes are small, shy creatures and, if left alone, will not cause harm. However, they will strike and bite when annoyed or handled without warning. They are mainly active at night, hiding during the day under rocks, logs and tree bark on the ground. They also hide under building materials and rubbish scattered in rural areas so be careful and wear gloves when removing objects that have been lying on the ground for some time.

This small snake is very difficult to identify and many cannot believe that a snake so small could have such a fatal bite. It is easily mistaken for a young red-bellied black snake or the harmless slaty-grey snake because they are all a similar colour.

WHERE TO WATCH OUT: Eastern small-eyed snakes live in a number of different habitats from Cape York to southern Victoria.

Blue-ringed octopus *Hapalochlaena maculosa*

When upset, bright blue warning rings flash over this octopus.

SIZE: Blue-ringed octopuses measure only about 10 centimetres from tentacle to tentacle.

NUMBER OF HUMAN DEATHS: Two people have died in Australia from this octopus's bite.

EFFECTS OF VENOM: The effects of a blue-ringed octopus bite happen quickly! Weakness and numbness around the face and neck is followed by breathing difficulty and nausea. A victim can die within 30 minutes from sudden paralysis or lack of oxygen. There is no antivenom. The only treatment is to perform *cardiopulmonary resuscitation* (CPR) for hours.

AVOIDING AN ATTACK: This colourful little octopus can be found stranded in pools or in crevices along rocky shores. It can change colour and shape to blend in with rocks or corals, so it may even look like a piece of rock. Because it is attractive and colourful, small children may be tempted to touch it if they find it in a rockpool. It will flash blue rings when touched or handled, but by then it is usually far too late! Never ever touch this octopus or poke it with your finger!

The blue-ringed octopus is the most dangerous octopus in the world. This tiny creature carries enough venom to kill 26 adults within minutes! Before it bites, this octopus flashes blue warning marks over its body and releases deadly saliva into its biting "beak", in the middle of its tentacles.

WHERE TO WATCH OUT: They live right around the Australian coastline, in rocky pools and crevices.

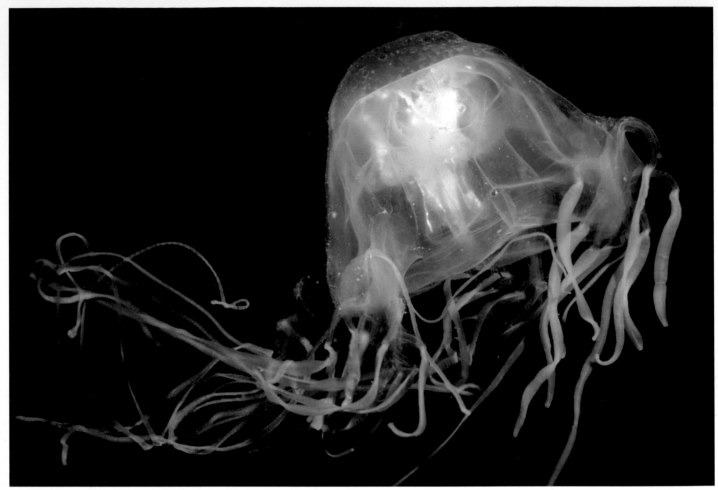

The box jelly has stinging, ribbon-like tentacles.

The box jelly is almost invisible in crystal clear water. Every summer, thousands of these deadly creatures invade the shallow waters of Australia's northern coastline.

SIZE: The box jelly's body (or bell) may measure 23 centimetres in length, but its tentacles can be many metres long. The box jelly can pull in its tentacles to avoid damage in rough, shallow water.

NUMBER OF HUMAN DEATHS: At least 60 people have died in Australia from box jelly stings.

EFFECTS OF VENOM: The box jelly has the most powerful, fast-acting venom in the world. The tentacles have millions of stinging cell organelles (extremely small organs inside the cell) called *nematocysts*. Each nematocyst leaves a red spot and a painful sting on any skin it comes into contact with, but because the stings are multiple, the red marks are shaped like whip lashes. The box jelly has 60 ribbon-like tentacles that can be several metres long. Even a mild sting can be extremely painful, so many of the box jelly's victims panic and drown. The venom can cause death within minutes. In fact, most victims died within 20 minutes of a sting. Although box jelly antivenom is available, the effects are so quick that it is often too late to use it.

AVOIDING CONTACT: Box jellies are most common in northern Australia during the warmer months from November to May, when swimming in the ocean is not recommended. Most tropical beaches have signs warning people to stay out of the sea and what to do if they are stung. Protective clothing, such as a wetsuit, prevents stings but the best prevention is not to swim in the sea during box jelly season.

Because of the severe pain, victims usually cannot apply any first aid, often they cannot even get out of the water without someone helping them! Once out of the water, lots of vinegar should be poured over the stung area for at least 30 seconds to prevent more nematocysts stinging. If the victim is unconscious, clear their airway and start resuscitation. If possible, a health professional should give the victim oxygen and antivenom. Fast transportation to the nearest hospital is crucial.

How bad and painful the sting is depends on where the victim is stung and over how much of the body. Small children are more at risk of dying from the box jelly's sting than adults.

WHERE TO WATCH OUT: The deadly box jelly frequents the shallow coastal waters off the tropical coast of northern Australia.

Irukandji *Carukia barnesi*

The irukandji is a very small but dangerous animal.

This tiny, see-through sea jelly leaves deadly nematocysts glued to the surface of its victim's skin. Nematocysts are tiny cells that explode and release venom when they touch skin or other objects. The internal parts of the cells are called cell organelles and inject venom into the victim's body.

SIZE: The body (or bell) is only 3 centimetres wide but the tentacles are up to 80 centimetres long.

NUMBER OF HUMAN DEATHS: There have been two deaths worldwide from irukandji sea jellies — both occurred in Australia.

EFFECTS OF STING: Victims first feel only mild pain but serious symptoms occur 20 minutes later. Small pimples on the surface of the skin appear and disappear within half an hour. After that, the victim suffers severe pain in the body or in the affected limb and joints, followed by nausea, vomiting, headaches and sweating. These symptoms usually ease after medical treatment. The sting is not always obvious but if suspected, victims should get out of the water fast!

AVOIDING CONTACT: The small size of this sea jelly makes its difficult to spot. Avoid swimming in areas where irukandji are suspected — that is the only real way to avoid a sting from these creatures.

WHERE TO WATCH OUT: Irukandji have been recorded around the Tropic of Capricorn and in northern Australia, but scientists suspect that these sea jellies may be spread throughout the world.

Reef stonefish *Synanceia verrucosa*

The reef stonefish has toxic spines along its back.

The reef stonefish is the most venomous fish on the planet! This strange-looking creature uses its large *pectoral fins* to bury itself in the sand, becoming almost invisible on the bottom with just its eyes and mouth sticking out. The stonefish prefers to live in shallow water, so most of its unsuspecting victims step on it while wading or swimming.

SIZE: Reef stonefish are 20–35 centimetres long.

NUMBER OF HUMAN DEATHS: The only reported Australian death is of an Army Medic who died a few days after a sting on Thursday Island in 1915.

EFFECTS OF VENOM: The pain caused by the spines piercing the flesh rapidly worsens and the foot swells up. The level of pain depends on how many spines have created the wound and how deep it is. The leg can become paralysed. The patient may also suffer from traumatic shock.

AVOIDING CONTACT: You don't have to be swimming in the ocean to step on a stonefish — they also live in shallow pools of water on the shore. Stonefish are not very good swimmers so they often get trapped in small pools when the tide goes out. Always wear solid shoes when wading in shallow water, and slowly shuffle your feet on the bottom.

WHAT TO DO IF STUNG: If stung, get to the hospital as quickly as possible. Immersing the wound in hot water helps relieve pain, but doctors can also give painkillers, such as morphine or pethidine, and can administer the antivenom.

White sharks have a high, pointed first *dorsal* fin that is curved at the tip and a pointed snout with a mouth full of razor-sharp teeth.

The white shark, sometimes called the "great white" or "white pointer" is the most ferocious animal in the sea! Its enormous mouth is full of large, widely-spaced, triangular-shaped teeth. It can easily bite a fully-grown seal in half or take large chunks out of an injured whale. Not only should divers and snorkellers be wary of a white shark swimming up from the depths; swimmers and surfers in shallow water can also be at risk. The white shark has a special way of moving blood through its body, keeping its body temperature warmer than the surrounding water. This gives it more energy and allows this shark to swim faster and further than other sharks on the same amount of food.

SIZE: The largest white shark measured 6 metres and weighed approximately 3000 kilograms.

NUMBER OF HUMAN DEATHS: White sharks are formidable predators that have carried out many fatal attacks. Worldwide, it is estimated that 63 people have been killed by this species of shark. Only a few lucky people have escaped the white shark's jaws to tell the tale. White sharks are large, carnivorous predators and regularly attack large, warm-blooded prey, so they are a threat to humans that come into their environment. Some scientists believe that most shark attacks occur when sharks confuse humans with seals or turtles.

AVOIDING AN ATTACK: A diver in a black wetsuit may look like a seal to a white shark. It is therefore very dangerous to swim, snorkel or dive in waters inhabited by seals. White sharks are also attracted by blood, so never swim or dive with any bleeding injuries or carry dead fish with you. Sharks have a very good sense of smell and can detect blood from a long way away. If you see a white shark, swim away from it smoothly, quickly and quietly. Splashing and panicking may make the shark think you are an injured animal — an easy meal for a shark!

Records show that most attacks take place in warmer months from November to March, within 200 metres from the shore in warm water near the surface — these are also the most popular places for humans to swim and snorkel. Most of Australia's popular beaches are now protected by shark nets. You should always swim between the flags where watchful lifeguards can reduce the risk of attack.

WHERE TO WATCH OUT: White sharks live in *temperate* and subtropical waters, mostly cooler than 20° Celsius. They are found from North West Cape in Western Australia, around the southern coast to Queensland, but are mostly seen in southern Australia.

White sharks make some spectacular attacks on prey, even launching out of the water to attack prey on the ocean surface.

White sharks are most often seen in shallow coastal waters, but they may also be found in the open ocean far from shore, and sometimes swim down to over 1000 metres. These fearsome predators are also found off New Zealand, South Africa and California, where they are especially common around seal and sea-lion colonies. White sharks can travel long distances, so they may occasionally inhabit temperate and subtropical seas throughout the world. White sharks are a protected species in Australia and many other parts of the world.

The white shark is the most feared of all sharks.

Sharp teeth and flexible jaws are perfect for tearing up large prey.

Tiger shark *Galeocerdo cuvier*

Tiger sharks have a blunt head and "tiger" stripes along the body.

The tiger shark is a dangerous predator with saw-edged teeth that cut and tear chunks from much larger animals, such as dead whales. This menacing creature is not a very fussy eater — plastic bottles, cans, rubber tyres, car number plates, glass bottles and turtle shells have been found in the stomachs of tiger sharks!

SIZE: Tiger sharks can grow up to 6 metres long.

NUMBER OF HUMAN DEATHS: Tiger sharks have been blamed for 84 deaths worldwide, but in some cases the species of shark may have been misidentified.

SEVERITY OF ATTACK: Tiger sharks are curious creatures, often swimming very close to divers to investigate. That curiosity, along with their vicious predatory habits, makes tiger sharks very dangerous. They can easily bite off a limb or kill a person.

AVOIDING AN ATTACK: Always pay attention to beach swimming rules and never swim in unprotected waters where tiger sharks have been reported. When fishing from a dinghy, never throw fish scraps overboard. Be careful when spearfishing — remember, sharks can smell blood over long distances!

WHERE TO WATCH OUT: Warm tropical waters are the preferred habitat for tiger sharks. They can be found in the shallows and even in muddy estuaries. They do, however, swim out further into 150-metre deep water in open oceans, even in temperate zones.

Bronze whaler *Carcharhinus brachyurus*

A bronzed sheen to this shark's skin led to the name bronze whaler.

The name bronze whaler shark is often incorrectly used for a number of different species of whaler sharks, which all have the same colouring and body shape.

SIZE: The bronze whaler shark grows to 3 metres long and can weigh 250–300 kilograms.

NUMBER OF HUMAN DEATHS: Bronze whalers are believed to have killed twelve people worldwide.

SEVERITY OF ATTACK: Cuts, tears, mauling, pain and shock can easily make a shark attack victim drown in the water while trying to escape the shark's jaws. The bronze whaler shark is not regarded as a man-eater, but the injuries it can cause may be fatal.

AVOIDING AN ATTACK: When hungry, the bronze whaler shark may attack swimmers and divers as well as surfers. Swimming, diving and surfing in murky waters is always risky. If you are visiting a new place and are not sure if it is safe to swim in the sea, always ask a local lifesaver first, before jumping into the water. Lifesavers will be able to tell you whether there have been recent sightings.

WHERE TO WATCH OUT: The bronze whaler shark is found all around the world, but in Australia it is only found in cool southern waters from south Queensland, around the southern shores to Jurien Bay in Western Australia. Bronze whaler sharks are also common around Moreton Island in Queensland.

Bull shark *Carcharhinus leucas*

Bull sharks often lurk in canals, harbours and rivers.

Bull sharks are smaller in size than white or tiger sharks, but they are just as dangerous! They are often found in areas where people swim and play. Unlike all other shark species, bull sharks can live for a long time in freshwater, and can even breed there. For this reason, they are sometimes also called "freshwater whalers" or "river whalers".

SIZE: Bull sharks grow to a maximum of 3.4 metres long. Females are larger than males.

NUMBER OF HUMAN DEATHS: This species of shark has killed seventeen people in unprovoked attacks throughout the world.

SEVERITY OF ATTACK: Bull sharks will eat anything that moves. Most attacks on people probably happen by mistake, because these sharks often swim in murky water where they might mistake a human for other prey. But even if they do make a mistake, once blood is in the water and there are other bull sharks around, the scene can easily turn into a feeding frenzy.

AVOIDING AN ATTACK: Never swim in murky water or in canals or harbours, and especially not at dusk or dawn.

WHERE TO WATCH OUT: Bull sharks live in warm tropical waters and are probably the most dangerous of all tropical sharks. Unlike most other shark species, bull sharks like to live in shallow estuaries and may even travel a long way up rivers into freshwater, where they swim in large groups.

Hammerhead sharks *Sphyrna* spp.

The great hammerhead is a huge shark that deserves respect.

All members of the hammerhead family have the unmistakable, hammer-shaped head that gives these predators their name.

SIZE: The great hammerhead, the largest of the family, can grow to a massive 6 metres long. Most grow to an average size of 2–3 metres.

NUMBER OF HUMAN DEATHS: Two deaths are said to have been caused by hammerheads worldwide, but in both cases there is some doubt as to whether a hammerhead was actually responsible.

SEVERITY OF ATTACK: Although they have small mouths, all species of hammerhead sharks have sharp teeth that can cause serious cuts. If the shark's teeth cut through an artery, a diver can easily bleed to death. Hammerheads are not aggressive sharks but any animal of that size, in or out of the water, should be respected.

AVOIDING AN ATTACK: Do not provoke a hammerhead shark. Unless angered, these sharks are not usually interested in divers or swimmers. If they come close to you, calmly swim away from them. However, hammerheads may become excited by spearfishing activity and blood in the water.

WHERE TO WATCH OUT: Hammerheads live in temperate and subtropical waters around the world. They are found from south-west Western Australia, and around the northern Australian coast to Victoria.

Cone shells *Conus* spp.

This textile cone shell's siphon is out, drawing water to its gills.

SIZE: Most species have a shell length of less than 10 centimetres.

NUMBER OF HUMAN DEATHS: Fifteen people are estimated to have died from cone shell stings. Only one death has been recorded in Australia (in 1935).

EFFECTS OF VENOM: A cone shell sting causes sharp pain, which may be excruciating. The strong neurotoxin causes other symptoms to develop quickly, including weakness, poor hearing and sight, and difficulty swallowing and speaking. Death is usually a result of paralysis of the lungs. Even in less severe cases, victims may take from a few hours to several weeks to fully recover from a cone shell sting.

There are more than 70 species of cone shells found in Australian waters. Five species of these marine snails are dangerous to humans. Cone shells are very slow-moving creatures, so they have to kill their prey quickly before it escapes. A barbed "harpoon", hidden in the cone shell's *proboscis,* shoots out to inject venom into the victim.

AVOIDING CONTACT: It is very easy to avoid contact with a cone shell — do not touch them! Even if a shell looks empty, the animal may be hidden deep inside. When exploring on coral reefs, always wear protective shoes.

WHERE TO WATCH OUT: Cone shells may be found in shallow water at low tide. They mostly live in warm, northern waters, particularly on coral reefs, around all but the southern coasts of Australia.

Poisonous crabs Various spp.

This small, bright crab is extremely poisonous.

SIZE: Sizes differ but all species are rather small.

NUMBER OF HUMAN DEATHS: No one has died from marine crab poisoning in Australia, although several people have died from eating these crustaceans in other parts of the world. It is also possible that deaths on isolated islands in the Pacific region were not reported or that the cause of death was uncertain.

EFFECTS OF POISON: The marine crab's poison is called saxitoxin. Saxitoxin slowly paralyses muscles throughout the body, eventually causing suffocation. There is enough of this poison in a single crab to kill thousands of humans. The poison dissolves in water and isn't ruined by boiling, which turns a soup of cooked crabs into a deadly broth.

Many species of marine crabs are poisonous when eaten. They are small and usually brightly coloured or with black-tipped claws. Most are easily found in shallow water. Other countries and tropical islands use a great variety of marine crabs for cooking; luckily, however, they are not eaten in Australia.

AVOIDING POISONING: Do not eat any small crabs, particularly brightly coloured species.

WHERE TO WATCH OUT: Avoid eating crab meat or soup unless you know what kind of crab was used in the dish. This warning applies when eating seafood in the Pacific Island region where people collect their seafood from the seashore rather than a shop.

A male Sydney funnel-web spider in a striking position.

No other spider in the world matches the male Sydney funnel-web's ability to kill humans. Although many mammals are not harmed by this spider's venom, it particularly affects the nervous system of humans and other *primates* and has caused a number of deaths in Australia. Surprisingly, funnel-web spiders from Sydney's north shore are twice as venomous as those found on the Central Coast of New South Wales. It is not known why the spider's location affects the strength of its venom. Although these spiders have a nasty reputation, they are not usually aggressive and will try to run away rather than attack.

The female funnel-web is larger than the male but less dangerous.

SIZE: The male spiders are smaller, measuring around 2.5 centimetres, while the larger females grow to 3.5 centimetres or more. Despite its smaller size, the male's venom is five times more dangerous than the female's, but both are a threat to humans.

NUMBER OF HUMAN DEATHS: Since the first death in 1927, thirteen people have died from a male Sydney funnel-web spider's bite. All deaths took place before the specific antivenom became available in 1980. Since then, there have been no more deaths.

EFFECTS OF VENOM: All bites must be treated as potentially life-threatening because the Sydney funnel-web's venom works very fast. Records show that five children under eight years of age have died within 90 minutes of the bite. The bite is painful, partly because of the large fangs and also because the venom is acidic. If first aid is not applied immediately, the symptoms can develop within ten minutes of the bite, starting with numbness around the mouth, spasms and twitching of the tongue. Vomiting, pain inside the body, salivation, heavy sweating followed by mental confusion and coma follow. Quick and correct first aid treatment is absolutely essential, as well as getting to the nearest hospital. Tests have shown that the Sydney funnel-web spider can deliver two effective bites one after the other, each time releasing the same amount of venom. The fangs may be stuck so deeply in the victim's skin that it can be difficult to remove the spider!

AVOIDING AN ATTACK: Most bites take place in early summer, when the male spiders are searching for females and often crawl into houses or sheds. These spiders prefer not to enter houses, but when they accidently do, they may not find their way out and end up hiding under objects left on the floor to avoid daylight.

If you are in funnel-web territory, shake out your clothes before putting them on, check your shoes and do not leave your clothes lying on the floor, even in the laundry. It is also a good idea to have regular pest control checks around and under the house. There are also three species of tree-dwelling funnel-web spiders, but the Sydney funnel-web is a ground spider.

WHERE TO WATCH OUT: Sydney funnel-web spiders don't just live in Sydney — they are also found around Nowra, Lithgow and Newcastle. Like most ground-dwelling spiders, this species lives in damp, dark burrows.

deadly

23

Red-back spider *Latrodectus hasselti*

The female red-back spider is larger than the male.

The red-back is perhaps the most famous of all Australian spiders. This small, soft-bodied spider killed thirteen people before an antivenom was developed in 1956. Only the female is dangerous to humans; the male's very short fangs are not able to pierce human skin.

SIZE: The adult female's body, which measures about 1 centimetre, is surrounded by long, slender legs. The male is much smaller, at around 3–4 millimetres.

NUMBER OF HUMAN DEATHS: Each year, almost 2000 people are bitten by red-back spiders, and of those, about 250 need antivenom treatment. Before the antivenom was developed in 1956, thirteen people died from red-back spider bites. Since then, only one person has died (in 1999).

EFFECTS OF VENOM: Being so small, the red-back spider is easily overlooked. Because the bite is not very painful or visible at first, it may be mistaken for a scratch and ignored. People bitten by a red-back spider usually feel a mild sting at the bite site, which becomes painful after 10–40 minutes. In more serious cases, nausea and vomiting, headache, fever, chest pain, muscle spasms and paralysis may follow.

AVOIDING AN ATTACK: When gardening or cleaning out the shed, wear gloves, a long sleeved shirt and closed footwear.

WHERE TO WATCH OUT: These spiders are found throughout Australia, including Tasmania.

Paralysis tick *Ixodes holocyclus*

Paralysis ticks crawl onto human skin and feast on their blood.

Although ticks are commonly found on dogs, cats and native mammals, humans can also be "hosts" to this nasty parasite. The biggest problem is that victims are often unaware that a tick is on them. Sickness and other symptoms may come as a surprise days later.

SIZE: An unfed adult female tick's body may be 3.8 millimetres long and 2.6 millimetres wide. But when full of blood, she can be 13.2 millimetres long and 10.2 millimetres wide!

NUMBER OF HUMAN DEATHS: At least 20 people died from a paralysis tick between 1904 and 1945.

EFFECTS OF VENOM: The venom is neurotoxic, causing muscle weakness, loss of muscle control and loss of balance. It may take up to three days for the tick to fill itself with enough blood. Afterwards the tick releases its toxic saliva into the wound. Some victims develop an allergy to the toxins and will have severe reactions if bitten by any ticks in the future.

AVOIDING AN ATTACK: Ticks are hard to avoid because they are very small. They are often not even felt at all until after they are firmly attached. Ticks can be picked up from grass or bushes just by brushing against vegetation. Attached ticks should be immediately removed, using a slim pair of tweezers or with fingernails. Be careful not to leave the tick's mouthparts embedded in the skin, as this may lead to a painful swelling of the affected area.

WHERE TO WATCH OUT: Paralysis ticks live within a 20-kilometre strip along the east coast of Australia, in moist forests, grassy areas and open forest areas where bandicoots, kangaroos, possums and humans live.

The giant bulldog ant is only found in Australia.

The group of insects known as bulldog ants are among the most primitive of all ants. When comparing them with fossils of their prehistoric ancestors, we can see how little they have evolved. Giant bulldog ants are the largest of the bulldog ant species and the largest ants in Australia. They are aggressive creatures that will attack if annoyed. Walking close to their nest can upset them, but damaging their nest in any way will most certainly cause an army of angry ants to rush out! With their *mandibles* wide apart, giant bulldog ants run quickly and fearlessly straight towards an intruder.

SIZE: Workers are between 1.3–3.6 centimetres long.

A giant bulldog ant nest may be up to 70 centimetres high.

Giant bulldog ants have long, fierce-looking jaws, called mandibles.

NUMBER OF HUMAN DEATHS: Several people have died in Australia as a result of a bulldog ant sting. The ants involved all belong to the genus *Myrmecia*, which is widespread throughout Australia. Most deaths were related to allergic reactions to the venom in their sting.

EFFECTS OF VENOM: Although the venom itself is not likely to kill a person, it can cause a strong allergy. An existing allergy can make a second sting very dangerous for the unlucky victim.

AVOIDING AN ATTACK: Although giant bulldog ants are large in the ant world, they can be hard to see in clumps of grass or in leaf litter. Their nests are well camouflaged, covered with dry leaves or pine needles, making them easy to bump into. Solid shoes, long pants and, most importantly, looking carefully before sitting down on the ground all reduce the risk of a bulldog ant attack. If confronted by angry bulldog ants, walk away quickly and then carefully check your clothing for crawling ants. A sting from a bulldog ant may be treated with antihistamine medication. Depending on the patient's reaction, application of antihistamine ointment, pills or injections may be necessary. It is best to seek medical advice.

WHERE TO WATCH OUT: This species is a tropical ant, found only in Australia. They build their nests in semi-dry forests, especially in casuarina forests. The nests are tall mounds, measuring up to 70 centimetres high. The mounds are usually covered in dry leaves or pine needles.

deadly

Wasps and bees *Hymenoptera* spp.

Most species of wasp are harmless to humans and important in the pollination of flowers.

There are hundreds of species of wasps and bees in Australia, but only a handful of them are dangerous, particularly to people who have an allergic reaction to the sting. Although none of the Australian species are aggressive, they will sting if you disturb their nests.

SIZE: Species vary between 1–3.5 centimetres long.

DANGER TO HUMANS: Generally, wasps and bees are not deadly to humans, but some people may have a severe allergic reaction to wasp or bee stings. Such people may die as a result of these stings.

AVOIDING AN ATTACK: Both wasps and bees are aggressive when their nests are disturbed. Some species swarm onto the intruder in massive numbers, causing many stings; other species attack alone. Some sting only when handled or stepped on. Wild wasps and bees should be avoided, especially by honey-bee keepers or those who have been bitten before. People who know they are definitely allergic to the sting should carry injectable adrenaline if they are likely to come into contact with bees or wasps.

WHERE TO WATCH OUT: Different wasps and bees build their nests in different places. Some live underground, others make nests from rolled up leaves, while some species make paper-like combs that hang from branches or buildings. Some wasps and bees are attracted to sweet foods and drinks. These should be covered by a cloth or net when eating outdoors.

Water buffalo *Bubalus bubalis*

Water buffalos are large, unpredictable animals.

Water buffalos are not native to Australia. They were brought in from Asia, where these massive mammals are *domesticated* and used as farm animals. Water buffalos are usually shy and timid; however, these large, horned beasts can sometimes be unpredictable, especially if wounded.

SIZE: An adult water buffalo stands about 1.8 metres high and can weigh up to 1180 kilograms.

SEVERITY OF ATTACK: The water buffalo is a strong beast that can move very quickly when it wants to. When aggressive, a water buffalo can gore a person with its horns or charge and stampede. A water buffalo has killed at least one person in Australia.

AVOIDING AN ATTACK: Never tease a buffalo! They will charge, even at a car. If on foot, run towards the nearest tree and climb up quickly!

Buffalos, as heavy and bulky as they are, can run very quickly and can move around a tree trunk with surprising speed and agility.

WHERE TO WATCH OUT: Water buffalos are mostly found only in Arnhem Land in the Northern Territory, where they live in swamps and along riverbanks.

They are good swimmers and can be seen crossing rivers or cooling off in billabongs. They are causing serious environmental damage in wetlands and are a declared pest.

deadly

deadly

Dingo *Canis lupus dingo*

Dingoes are pack animals and powerful carnivores, but in the wild they usually flee from humans.

The dingo is a wild dog that was brought to the Australian continent by Asian seafarers about 4000 years ago. It is a shy animal that lives in small territorial packs. Over time, the dingo has interbred with domestic dogs, so there are few purebred dingoes remaining today. Fraser Island in Queensland, is home to one of only a few colonies of purebred dingoes. Dingoes, like many dog species, can be potentially dangerous in close encounters with humans.

SIZE: An adult dingo stands about 60 centimetres tall at the shoulder and is about the size of a large dog. Males are generally slightly larger than females.

SEVERITY OF ATTACK: The dingo is the largest *carnivore* in Australia. There have been two reports of fatal dingo attacks on humans, one of which was the death of a nine-year-old boy on Fraser Island in 2001.

AVOIDING AN ATTACK Under normal circumstances, dingoes do not attack humans. They mostly do their best to avoid people. However, in some popular tourist areas, dingoes that have been offered food in the past have become used to humans and may act aggressively in an attempt to get food. Never approach a dingo or offer it something to eat.

WHERE TO WATCH OUT: There are now warning signs at some popular picnic and camping grounds where these dogs have become used to people and now expect to be fed. When visiting these areas, do not leave any food out in the open, especially overnight, and never feed dingoes. Avoid wandering away by yourself because a dingo will more likely attack if you are alone. Look after children at all times.

Fraser Island is one of few places where purebred dingoes survive.

The most familiar colour of a dingo is ginger, but they may also be black, white or tan coloured.

deadly

27

deadly

Southern cassowary *Casuarius casuarius*

The southern cassowary is generally a shy bird.

People lucky enough to see an adult cassowary in the rainforest are often amazed by its huge size and stunning appearance. The cassowary's beak, and fearsome-looking "helmet", or casque, are harmless and never used as weapons. Instead, a cassowary uses its powerful legs equipped with dagger-like claws to kick and defend itself.

SIZE: An adult southern cassowary stands about 1.75 metres tall. Females are larger than males and have a taller casque.

NUMBER OF HUMAN DEATHS: In 1926, a sixteen year old boy from Mossman in north Queensland died — the result of a cassowary kicking him in the throat and slashing his jugular vein. Reports say that the boy and his dog had been chasing the cassowary.

SEVERITY OF ATTACK: A cassowary won't attack a person unless harassed, surprised or protecting its chicks. At close range, a cassowary can jump high into the air and kick with one or both of its strong legs.

AVOIDING AN ATTACK: Never run away from a cassowary! These birds can run much faster than humans and they seem to enjoy the chase. If you meet a cassowary in the wild, stand still and wait for the bird to pass. In the event of an attack, stand tall and raise your arms above your head. If you are with a group of people, bunch up and shake your arms in the air.

WHERE TO WATCH OUT: Cassowaries live in small areas of tropical rainforest in north Queensland.

Southern eagle ray *Myliobatis australis*

Because of its beautiful patterns, this ray was once known as the blue-spotted eagle ray.

These graceful swimmers are harmless to humans unless stepped on or handled. The whip-like tail has a sharp spine near its end, which is connected to tissue that produces venom. The spine is covered by a thin sheath, which also releases venom. When stepped upon or speared from a close distance, the stingray whips the spine into the victim's body, causing serious injury.

SIZE: This ray measures 1.2 metres across the body and grows up to 2.4 metres long, including the tail.

SEVERITY OF ATTACK: The tail spine, also called a "barb", may dig deep into the victim's flesh, causing a painful puncture wound. The tip of the spine can also break off, and stay stuck in the wound. Severe pain and trauma are the main results of a stingray injury but sometimes nausea, salivation, vomiting, sweating and muscle cramps may also follow.

In the most severe incidents, a ray's barb may puncture a person's heart. This causes such a massive shock to the system that the victim suffers a heart attack. World-famous Australian environmentalist, Steve Irwin, died tragically in waters off the north Queensland coast in 2006, the result of being whacked in the chest by the spine of a bull ray he was swimming above.

AVOIDING CONTACT: Stingrays are hard to see when resting on the sandy sea floor. They may also be dangerous if caught on the end of a fishing line. It's better to cut the line than try to remove the hook.

WHERE TO WATCH OUT: From Jurien Bay in Western Australia, along the southern coast (including Tasmania) and up to Moreton Bay in Queensland.

An olive sea snake has a paddle-shaped tail and is an excellent swimmer that can hold its breath for up to half an hour!

The olive sea snake is the most common species of sea snake encountered by divers. Unlike its close relatives, this marine species has fangs long enough (3–4.7 millimetres) to bite through a diver's wetsuit. These curious snakes often swim in for a closer look at divers; however, they are usually not aggressive unless deliberately annoyed.

SIZE: Olive sea snakes may reach 2 metres in length.

EFFECTS OF VENOM: The venom of this sea snake is certainly strong enough to kill a person. Apart from a mild jab, there may be little pain and the fang marks can barely be seen. In rare cases, the venom's effect may not be obvious immediately, but in most cases the affected person goes into shock very quickly (within 15–30 minutes after being bitten), experiences nausea and vomiting and may then collapse.

The venom is neurotoxic, attacking the nervous system. It also contains a chemical called myotoxin, which affects muscles. Victims can drown, especially if they are unaware of the bite or they ignore it. Bites from this species, however, are rare and no one has died from a sea snake bite in Australia.

AVOIDING AN ATTACK: Unlike venomous land snakes, a sea snake cannot open its mouth very wide, which means that it cannot bite easily. If you come face to face with an olive sea snake, stay still and calm. They are curious, non-aggressive creatures that will soon lose interest and swim away.

WHERE TO WATCH OUT: Olive sea snakes are found in tropical coastal waters in northern Australia, and New Guinea, mainly on coral reefs.

deadly

Pufferfishes *Arothron meleagris*

The skin and internal organs of the whitespotted pufferfish, and that of other pufferfishes, are very poisonous.

Pufferfishes are very poisonous. In Japan, eating them is considered a special treat because it gives a pleasant tingling sensation on the lips. But the risk of poisoning from eating pufferfishes is high and many people have died, even when the meal has been prepared by specially trained "fugu" chefs.

SIZE: Pufferfishes range from 15–76 centimetres long when relaxed, but can puff themselves up to the size of a soccer ball.

EFFECTS OF POISONING: The flesh and internal organs of pufferfishes contain tetrodoxin, one of the most powerful poisons in the world! Pufferfish poisoning is life-threatening. Its effects may be felt in as little as ten minutes and last up to five days. The affected person often feels tingling sensations and nausea but may not vomit. Some people experience numbness, paralysed limbs, an inability to breathe, dizziness and dilated (or wide) pupils.

First aid for tetrodoxin poisoning is limited. A poisoned victim should be made to vomit and taken to hospital.

AVOIDING POISONING: Do not eat pufferfishes or even allow any body fluids to squirt onto your skin!

WHERE TO WATCH OUT: Pufferfishes are common right around Australia's coasts, usually in very shallow water. They are often caught by anglers and are easy to recognise. While still on the hook they usually begin inflating themselves and should be quickly released.

deadly

Common lionfish *Pterois volitans*

The common lionfish's bright colours and patterns warn predators that this fish is dangerous!

The common lionfish, also known as the "butterfly cod", is a very beautiful and graceful fish. Because they are curious, these fish often swim towards divers, usually in pairs. They are also popular as aquarium pets. Owners often do not realise the potential danger these delicate creatures can cause. The common lionfish has thirteen dorsal spines, three anal spines and two *pelvic spines* arranged between its fins. Each spine is attached to a pair of small venom glands that pump venom to the tip of the spine when pressure is applied. The spine tips often break off on contact, causing a painful reaction on the skin. Common lionfish do not try to attack humans and injuries are rare. There are similar venomous fishes in our oceans, called "scorpionfishes". They look different but have similar venomous spines.

SIZE: Common lionfish grow to 30 centimetres long.

EFFECTS OF VENOM: The venom from a common lionfish is very similar to that of a stonefish. The sting is painful and becomes more painful over a few minutes. The excruciating pain usually disappears after a few hours but may last for several days. At the site where the venomous spines puncture the skin, slight swelling and numbness occur, and the surrounding area turns blue. The degree of pain and other health problems depends on how many spines jab into the skin and how deep they penetrate. In severe cases, the injured person may sweat a lot, vomit and have a fever.

Although there is no record of a person dying from a common lionfish injury in Australia, deaths have been recorded in other countries.

AVOIDING AN ATTACK: The striking red or brown bars on its body, fins and spines should make the common lionfish easy to spot. However, the fish blends in with colourful corals in the flickering light on the reef. Such bright colours, patterns and shapes are there for a good reason — the lionfish is warning other creatures not to go near it, and this warning should always be respected.

To avoid an injury, do not touch the lionfish or its spines, not even with gloves on! If stung, placing the wounded area in hot water can help the pain subside.

WHERE TO WATCH OUT: The common lionfish can be found in tropical and subtropical waters around Australia, especially on coral reefs, where the lionfish is best camouflaged against colourful rocks and corals. If you have a common lionfish in your marine aquarium, or are thinking of getting one, be careful not to touch your beautiful pet with bare hands.

Wolf spiders *Lycosa* spp.

Wolf spiders have excellent eyesight and remember landmarks.

Wolf spiders are very common and can be found in the bush as well as in city parks, median strips and backyards. They are good hunters and chase their prey quickly, holding it down with their fangs.

SIZE: Size depends on the species, but most wolf spiders are about the size of a fifty-cent coin.

EFFECTS OF VENOM: The bite from a wolf spider is not life-threatening but can cause a reaction in the area of the sting. When the spider bites, it leaves digestive juices on the skin at the bite site, which causes skin ulceration and lesions. The damaged skin and underlying tissues are not easy to treat and take a long time to heal.

AVOIDING AN ATTACK: Wearing shoes at night is enough to prevent a bite on the foot, but remember to look before you sit down on the ground. A wolf spider's eyes reflect brightly in torchlight, so it is easy to spot them at night as they hunt.

WHERE TO WATCH OUT: There are about 120 wolf spider species in Australia. They live in all kinds of habitats throughout the continent and are ground-dwellers, hiding in burrows during the day and hunting at night. Wolf spiders have excellent eyesight (they have eight eyes) and they can recognise and remember landmarks used to find their way around. Wolf spiders love manicured lawns on golf courses, bowling greens and in parks, where they are commonly found.

dangerous

White-tailed spiders *Lampona* spp.

Most white-tailed spider bites occur inside people's homes.

The white-tailed spider has a white spot at the tip of the abdomen. The two most commonly seen species are the southern species, *Lampona cylindrata*, and the similar looking eastern species, *Lampona murina*.

SIZE: This spider has thin, pin-sized legs but its size is about the same as a fifty-cent coin.

EFFECTS OF VENOM: Victims have complained of intense pain and itchiness where bitten, and swelling that often lasts much longer than the bite of other spider species. Despite popular belief, scientific evidence shows that a bite from the white-tailed spider does not result in severe ulceration and damage to body tissue, called necrosis. Ulceration is uncommon, and is probably the result of an infection.

AVOIDING AN ATTACK: White-tailed spiders may crawl into people's houses in search of other spiders to eat. One of the easiest ways to avoid this is to keep your home free of spider webs and therefore spiders. If a white tailed spider is found inside the house, you can get an adult to try to gently (and carefully) remove it with a broom or dust pan.

WHERE TO WATCH OUT: Almost all white-tailed spider bites occur indoors, mostly in bathrooms and bedrooms. They tend to hide in dark places, behind pictures on the wall, furniture and beds. Most reported bites occurred in southern Australian States.

dangerous

Whistling spider *Selenocosmia* sp.

The whistling spider is Australia's largest spider.

The whistling spider is the biggest Australian spider and it is very aggressive when it is disturbed. Its name comes from the sounds it makes. The spider rubs the base of its front legs (called *pedipalps*) against the base of its fangs. Patches of flattened spines are rubbed up and down, making a noise that sounds like whistling.

SIZE: These spiders can grow larger than a human handspan. Spiders are difficult to measure — the legs are longer than the body, but in a natural position their legs are never fully stretched out.

EFFECTS OF VENOM: Whistling spiders have very long fangs and are venomous but are not considered deadly to humans. However, being bitten by such a large spider is very painful and it could cause nausea and vomiting. There is a small chance that later on the area of the bite may blister and this wound may take some time to heal.

AVOIDING AN ATTACK: These spiders are ground-dwelling creatures that are active at night. They cannot jump, but they can move very fast, lifting their front legs up into the air and aiming their fangs at any threats. In this position, they can only strike downwards, not really forwards. Whether they are aggressive or not, they should be left alone.

WHERE TO WATCH OUT: Whistling spiders are found in north and western Queensland and New South Wales. Be careful when lifting logs or other objects off the ground — take care that there are no spiders lurking underneath!

Red-bellied black snake *Pseudechis porphyriacus*

Red-bellied black snakes often live close to water and prey on frogs.

This beautiful snake is also known as the "common black snake" and is probably the best known of Australia's venomous snakes. Because it lives in many places in eastern and southern parts of the continent, lots of people come in contact with the red-bellied black snake, but it rarely bites.

SIZE: This large, stocky snake reaches up to 2 metres in length, although is normally around 1.5 metres.

EFFECTS OF VENOM: The red-bellied black snake's venom is not very strong and it also only has short fangs. There is usually mild pain near the bite, but in more serious cases, nausea, vomiting, headache and chest pains can develop soon after. The venom contains neurotoxins, myotoxins, which affect the muscles and kidneys, and properties that prevent the blood from clotting. If bitten, people should seek medical attention as tiger snake or black snake antivenom is available and may be necessary.

AVOIDING AN ATTACK: Watch where you step, look down before you sit on the ground and don't run fast through high grass near water. Red-bellied black snakes are shy and will move away at the first sign of approaching danger.

WHERE TO WATCH OUT: Red-bellied black snakes live in grasslands and forests, but prefer moist habitats of New South Wales, Victoria, and small parts of South Australia and Queensland. They are not found in Tasmania, Western Australia, or the Northern Territory.

Copperhead *Austrelaps superbus*

Copperheads live in lowlands as well as in high mountains.

This *diurnal* snake is venomous enough to kill a person but is far from aggressive. Instead, the copperhead is a shy, timid snake that likes to shelter under logs, rocks and even sheets of iron. If disturbed, a copperhead prefers to try and escape rather than bite.

SIZE: The average size of an adult is 1.7 metres.

EFFECTS OF VENOM: A copperhead's venom is similar to that of a tiger snake but much weaker. A copperhead may inject large amounts of venom through its short fangs. A bite from this snake must be treated with caution. The venom is neurotoxic, so it can work rapidly and have serious effects.

AVOIDING AN ATTACK: Copperheads are very shy snakes but because they love the early morning sun after a freezing night, they can be a surprise for any early risers. They like to bask between large bunches of grass that protect them from strong winds and also hide the snake. When cornered, a copperhead will flatten its neck, hiss and take up an aggressive pose. Copperheads rarely bite and if they do, their short fangs cannot penetrate sturdy boots.

WHERE TO WATCH OUT: Copperheads live near swamps, creeks and moist places where they feed mainly on frogs. They are found in south-eastern Australia, from the western plains of New South Wales to the South Australian borders, as well as in Tasmania.

Collett's snake *Pseudechis colletti*

This snake has beautiful colours and markings, but is rarely seen.

Collett's snake is also known as Down's tiger snake. It belongs to the black snake group along with four other species, all of which are dangerous. This snake is the most colourful and the least dangerous of the five species. Because of its striking looks, Collett's snake is popular with collectors.

SIZE: Collett's snakes grow to about 1.5 metres long.

EFFECTS OF VENOM: Symptoms at the bite site include painful swelling, bruising and sometimes cell damage in the surrounding skin. Similar to a bite from its close relative, the red-bellied black snake, victims may also experience muscle and kidney damage. It is recommended that black snake antivenom be used in the treatment of bites by Collett's snake.

AVOIDING AN ATTACK: These snakes are rare and quite difficult to find. Even experienced herpetologists (reptile experts) have to search for days to find a Collett's snake in the wild.

During floods, these snakes seek high ground such as railway lines to escape the flooding waters, but even then they do not attack people — they just want to get away from them.

WHERE TO WATCH OUT: Collett's snakes live in central Queensland in areas that are not frequently visited by humans. So the most likely encounter you will have with a Collett's snake is if you have a friend who is a herpetologist or reptile keeper. Like most snakes, Collett's snakes do not like to be handled and they can bite if they feel threatened.

Freshwater crocodile *Crocodylus johnstoni*

Unlike its frightening cousin, the smaller freshwater crocodile is not a danger to humans unless it is harassed. This species of crocodile has a narrower snout and sharp, evenly-sized teeth. Freshwater crocodiles are found only in Australia.

SIZE: This species is smaller than the saltwater crocodile and grows up to 3 metres.

SEVERITY OF ATTACK: Only a few people have been bitten by freshwater crocodiles, all of them while handling or trying to catch the croc!

AVOIDING AN ATTACK: These crocodiles are shy, timid animals, hiding on the bottom when disturbed by humans. Freshwater crocodiles are harmless fish-eaters that only attack when they are deliberately annoyed by people.

WHERE TO WATCH OUT: Freshwater crocodiles are found inland across the top of Australia, in permanent freshwater areas such as rivers, creeks and lagoons. They spend most of their time in the water but will climb onto the banks to lie in the sun.

The freshwater crocodile has a narrow snout and evenly-sized teeth.

Platypus *Ornithorhynchus anatinus*

Only the male platypus has a sharp, 1.5-centimetre-long venomous *spur* on each ankle.

The male platypus is Australia's only venomous mammal. Females are harmless but males have a venomous spur which they use during the breeding season to fight with other males. Positioned on each ankle, this sharp, hollow spur is connected by a venom duct to a venom gland in each thigh.

SIZE: A mature male platypus averages about 50 centimetres in length. Females are slightly smaller.

EFFECTS OF VENOM: If scratched by a male platypus's venomous spur, the wound will be extremely painful for the first 30 minutes. The affected area may become numb and swollen. There is no specific first aid procedure, but an injury from a platypus is not life threatening.

AVOIDING AN ATTACK: Platypus injuries mostly occur when the animal is handled. They are usually shy, secretive creatures, but may sometimes become tangled in fishing nets or yabby pots and need to be released. Other than on these occasions, there is no reason to even touch a platypus. The male's spur is about 1.5 centimetres long, so it can penetrate through a towel or gloves. Although a scratch from a male's spur is capable of inflicting pain at any time throughout the year, the venom glands become more active during the breeding season, which differs between southern and northern populations.

WHERE TO WATCH OUT: Platypuses are very sensitive and only live in unpolluted freshwater creeks, rivers and lagoons in eastern Australia and Tasmania.

Flower sea urchin *Toxopneustes pileolus*

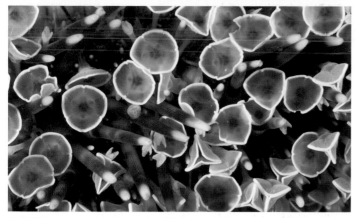

Each "flower" on this sea urchin hides a set of sharp jaws that contain sacs of chemicals toxic enough to kill fish and even humans!

There are many harmful sea urchins in the sea but this is one of the most dangerous. First discovered in 1961 in Moreton Bay in Queensland, it is common throughout the Indo-Pacific area, including Japan. The cup-shaped ends of the spines look like beautiful blossoms, but if you see one, don't touch it!

SIZE: The average size of a flower sea urchin is 9.5 centimetres with spines measuring just 8 millimetres.

EFFECTS OF VENOM: The parts of the urchin that look like flowers are called *pedicellariae* and they are used to catch the urchin's prey. Every pedicellaria has three tiny, sharp jaws, each attached to a venom gland. While the spines are not venomous, touching them releases the venom from the glands and causes severe pain that lasts for nearly an hour. In more serious cases, the victim may collapse and become paralysed, leading to breathing difficulties and even death. Although there are reports of Japanese divers dying from touching the flower sea urchin, there is nothing to confirm these reports. Flower sea urchins are also poisonous when eaten.

AVOIDING CONTACT: Despite its beauty, this flower-like animal should never be touched with a bare hand or exposed skin.

WHERE TO WATCH OUT: Flower sea urchins are widespread. This particular species is found along the east coast of Australia.

Kangaroos *Macropus* spp.

Kangaroos can kick with both back legs at the same time and "box" and scratch with their front legs.

Kangaroos are shy and timid in the wild, but in captivity or when injured, they can cause injuries with their powerful hind legs. When larger species, such as the red kangaroo, wallaroo, eastern or western grey kangaroo stand up on their hind legs, they can be as tall as an adult human.

SIZE: Some males can grow to almost 2 metres, including the tail.

SEVERITY OF ATTACK: Kangaroos can scratch and box with their front paws, which have long, sharp claws. A kick from the powerful hind legs of a large kangaroo can break a person's ribs, or cause groin injury, bruising, and even *disembowel* a person.

AVOIDING AN ATTACK: There is little risk of a kangaroo attack in the wild, unless the animal is injured. If you ever find a large kangaroo tangled up in a fencing wire or hit by a car and still alive, be very careful, as they also have sharp teeth and are capable of biting.

Many zoos have open areas where people can feed and pat kangaroos. In these situations, sometimes a kangaroo may push a child over or accidentally scratch with its front paws in an attempt to get more food. If a kangaroo starts to cough or rear up, walk away from it!

WHERE TO WATCH OUT: Helping injured kangaroos is a kind thing to do, but it is better if professionals help them because they know what they are doing. Never tease kangaroos in wildlife parks. If they want to just lie around or sleep, let them!

Cane toads are the largest amphibians in Australia.

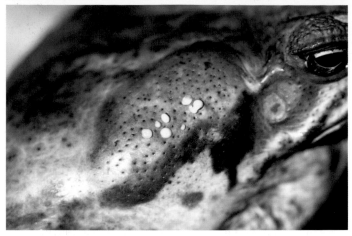

Milky poison oozes from the gland behind the cane toad's ear.

The cane toad was brought to Australia in 1935 to control the introduced cane beetle, a pest that damaged sugar cane crops. Unfortunately, the release of the cane toad in Queensland was one of the worst pest control disasters in history! As it turned out, the cane toads did eat some cane beetles, but also developed a taste for many other things, including native animals. Not only are they fierce predators, cane toads are highly poisonous to most animals that try to eat them. Cane toads are able to live in all kinds of habitats, and are breeding and spreading further across the country each year.

Stay away from the poisonous glands near the toad's head!

SIZE: Cane toads are the largest amphibians in Australia. On average, an adult cane toad is 10–15 centimetres long, but some females may reach an incredible 20 centimetres!

EFFECTS OF VENOM: Venom is produced in a large gland just behind each eye and when cane toads are handled venom may ooze onto the surface of the glands. The venom from a cane toad is called bufotoxin. Although no one has died from cane toad poisoning, large doses of this toxin may cause serious illness. But it's not only the adults that are dangerous — even a cane toad's tadpoles are poisonous!

Cane toads lay very distinctive, shiny black eggs.

AVOIDING CONTACT: If you must handle a cane toad, always wash your hands afterwards, and never touch or rub your eyes!

Although humans don't eat cane toads, if a person licks his/her fingers after handling a cane toad, the venom can cause serious illness.

Giant cane toads can grow up to 20 centimetres long!

WHERE TO WATCH OUT: Cane toads are ground-dwellers. They cannot climb and are generally quite clumsy. Their range extends throughout eastern and northern Queensland, across into parts of the Northern Territory and down into northern New South Wales, with an isolated *colony* surviving at Port Macquarie.

Port Jackson shark *Heterodontus portjacksoni*

The Port Jackson shark has a distinctive, "harness-like" pattern of darker bars on its body and fins.

The Port Jackson shark has a strong, sharp spine in front of each of its two, triangular dorsal fins. These spines help protect the shark against larger predators. Port Jackson sharks are living representatives of a long *extinct* group of sharks known only from 280–385 million-year-old fossils.

SIZE: Adults measure about 1.4 metres.

EFFECTS OF VENOM: The Port Jackson shark has a venomous spine in front of each of its two dorsal fins. When speared or caught on a hook, the shark thrashes around violently, and if close enough, a person may be scratched or stabbed by these spines. Apart from wounds, the spines contain venom that may cause muscle weakness, which can last for many hours.

Injuries from this shark are not common; it is a creature that minds its own business when left alone. It is recommended to treat the wound as you would an ordinary cut or scratch.

AVOIDING AN ATTACK: By avoiding direct contact with the shark, injuries should not occur. If a Port Jackson shark is hooked on a fishing line, cut the line and let the shark go rather than try to free the hook from the shark's mouth. Always be careful to stay clear of the shark's dorsal fins!

WHERE TO WATCH OUT: These sharks are found all around southern Australia, between Carnarvon in Western Australia and Byron Bay in New South Wales.

Striped catfish *Plotosus lineatus*

A large school of striped catfish look spectacular in the water.

There are more than 30 species of catfish in Australia, and all are able to cause injury with their venomous spines. Striped catfish swim in large schools and are easily identified by their white body stripes. When threatened by a predator, they raise their venomous spines, which makes them very difficult to swallow.

SIZE: The average size of the striped catfish is 35 centimetres, but some grow up to 1 metre long.

SEVERITY OF INJURY: Catfishes have one dorsal and two pectoral spines, which are very sharp. They also have two thin venom glands running along the spine. When the spine penetrates the flesh, the venom gland breaks, releasing the venom into the wound.

It takes a long time for the venom to break down, so treading on a dead catfish can be just as painful as treading on a live one! The pain usually does not last more than a couple of hours but the wounds may take several weeks to heal.

AVOIDING CONTACT: Catfishes are not dangerous while swimming in open water, as they tend to stay away from swimmers. However, striped catfish often become tangled in fishing nets and are hard to remove. Be very careful while untangling a slippery, thrashing catfish. Rather than getting too close, it is best to try to shake the fish out of the net instead.

WHERE TO WATCH OUT: Striped catfish live in estuaries along the northern coasts of Australia.

Flatheads *Platycephalus* spp.

Flatheads have excellent camouflage as they lie on the sea floor.

The flathead is a good eating fish and is commonly caught on fishing lines. There are more than 30 species of flathead found in Australian waters. They all look similar, with a flat body and head. Their bulging eyes are positioned on top of the head, making it easier to spy predators above.

SIZE: Depending on the species, flatheads average about 50 centimetres in length but some are over 1 metre long.

SEVERITY OF INJURY: The dangerous spines are not the long ones on the fish's back but the ones on each side, attached near the gill slits. If threatened, the fish quickly raises the spines upwards like a row of sharp spears. The dorsal spines are not venomous but they can cause a lot of pain and can frighten fishers who are not careful when handling the flathead. There is always a risk of infection, so the wound should be cleaned and treated like any other type of cut or scratch.

AVOIDING AN ATTACK: Be careful when taking a flathead off the hook. They are slippery, slimy fish, so it is best to cover the fish with a thick, wet cloth before grabbing it with your hands. Even then, the sharp spines can pierce the cloth and push through into the skin. Use pliers to remove the hook.

WHERE TO WATCH OUT: Flatheads are found from the south-west Western Australian coast around to south-east Queensland.

Portuguese man-o-war *Physalia physalis*

Stinging tentacles dangle from the Portuguese man-o-war's float.

The Potuguese man-o-war, or "bluebottle", is not a single animal but actually a floating colony of animals known as polyps. The gas-filled float is one polyp, while the animals making up the tentacles are specialised to catch prey. Together, they drift wherever the currents and wind takes them, stinging plankton, trapping fishes and eating them on their way.

SIZE: The float measures 2–13 centimetres and the trailing, frilled tentacles may be up to 10 metres long.

EFFECTS OF VENOM: Portuguese man-o-wars have stinging cells, called nematocysts on their tentacles. The nematocysts shoot barbed stings loaded with venom into the victim's skin when touched. The sting causes sharp pain, which becomes worse shortly afterwards. The stung area becomes red with tiny welts but the pain usually fades within 24 hours. No one has died from the sting in Australia, but several people have died in America from a similar species that lives in the Atlantic Ocean.

AVOIDING CONTACT: If there are bluebottles on the beach, there is a good chance that there are some in the water as well. If stung, pick off the tentacles and put an ice pack onto the stung area to ease the pain. Vinegar also helps to destroy the poison.

WHERE TO WATCH OUT: Bluebottles are found in waters right around Australia but are more common in the summer months. They can be near the beaches in very large numbers, especially on a windy day.

Fireweed *Lytocarpus phillippinue* or *Algaophenia cipressina*

The tiny animals living together are extremely hard to see!

The fireweed is not actually a plant but a colony of extremely tiny animals living together as one, generally called a "hydroid". The two species of fireweed are very similar. They are colonies of very small creatures that form feathery *fronds* containing thousands of nematocysts. These tiny cells release venom when touched, causing severe pain.

SIZE: The size differs greatly, depending on the hydroid's age and position on the reef or rocks. They are usually about the size of a small garden fern.

EFFECTS OF VENOM: Accidents mostly happen when a diver or snorkeller brushes against the hydroid with bare skin. The disturbed colony releases its stinging chemical into the water, so other swimmers passing close by may also be stung. The itching pain is immediate and may increase over time. A rash and blisters may develop on the stung skin and in some cases the victim may suffer a fever and general sickness. A hydroid sting is not life threatening and usually does not require medical treatment.

AVOIDING CONTACT: When swimming, diving or snorkelling, stay clear of these fern-like hydroids. Their stings can't penetrate through a wetsuit or any other protective clothing, so wearing these may prevent an unpleasant experience with the hydroid.

WHERE TO WATCH OUT: Both species are most common in warm waters of northern Australia, always below the low tide water mark. Hydroids can also be attached to submerged parts of wharfs, pylons and other artificial structures.

harmful

Land leeches Various spp.

A leech sucks blood through a mouth at one end of its body.

Land leeches are parasites that feed on blood. They use their three-toothed jaws to cut a circular hole into the skin of a "blood donor". At the same time, they inject a chemical called hirudin into the wound. Hirudin prevents the blood from clotting once inside the leech, so it can be digested. Hirudin also acts as an anaesthetic, making the bite painless. Once a land leech is full of blood, it drops off, leaving the wound bleeding for a long time, even several hours.

SIZE: The size of a leech depends on whether it is stretched or shrivelled. Stretched, some leeches can be up to 15 centimetres long!

SEVERITY OF BITE: Leeches do not transmit diseases but the bacteria in their guts can cause an infection in the wound. After the leech has dropped off, the prolonged bleeding, although gory to look at, is never a serious problem, even if several leeches were sucking at the same time. Some people experience slight swelling and itchiness for several days after a bite. Leech bites do not require medical help.

AVOIDING CONTACT: It is best to avoid walking in leech-infested areas, especially on rainy days. Insect repellents do work for a little while but do not work well in wet weather. There are many ways to remove a leech, such as burning it with a flame, sprinkling salt onto it, or leaving it there till it drops off, but the simplest way to remove a leech is to pull it off, flick it far away and then rub styptic pencil (Alum) over the wound to stop the bleeding.

WHERE TO WATCH OUT: Land leeches mainly live in rainforests, and in moist parts of drier forests.

harmful

Centipede *Scolopendromorpha* sp.

Centipedes do not attack humans but may accidentally crawl onto a person who is lying on the ground.

The centipede has a pair of large, powerful fangs at the front of its head. As well as having an unpleasant bite, a centipede is hard to remove. Each of its many legs ends with a sharp claw and digs into the skin while the jaws are still fastened. If you try to pull the centipede off, chances are that you will get bitten again — on the finger!

SIZE: Centipedes grow mostly to around 10 centimetres but some species grow twice that size.

EFFECTS OF VENOM: The bite from a centipede is extremely painful and the pain may last for 24–48 hours. The venom may cause some mild skin damage where bitten, but there is usually no need for medical treatment.

As with all bites, cuts or scratches, there is a possibility of a secondary infection, whereby bacteria are introduced into the victim at the site of the bite. Although there is no prescribed first aid, one way of relieving the pain is to run hot water over the site of the bite.

AVOIDING CONTACT: Centipedes are nocturnal predators. They have poor vision and are easily stepped on. Do not walk around without shoes on in the bush at any time, particularly at night. Do not poke a centipede with a stick, as it may quickly crawl onto the stick and onto your hand!

WHERE TO WATCH OUT: Centipedes live underneath logs, rocks or bark. If there is a centipede crawling on your clothing, try to quickly flick it off with a stick, never with your hand!

Scorpion *Liocheles* sp.

Scorpions often prey upon spiders. They use their strong front claws to hold prey tightly while the stinging tail paralyses the victim.

While many people die from scorpion stings each year in other countries around the world, Australian scorpions are not deadly to humans. They are mostly small and their venom is weak. There are many species in Australia and their classification is not well understood because these creatures haven't been studied much.

SIZE: Most scorpions are around 7 centimetres long, including the tail.

EFFECTS OF VENOM: Scorpions have two claws, which are used to grab unsuspecting prey. The scorpion then drives the sting, which is on the end of its tail, into the victim. Scorpions only sting people when handled or harassed. The weak venom makes the sting painful but.only for about 15–45 minutes. There is no need for any medical treatment, unless a secondary infection sets in, the result of bacteria being pushed into the wound. This may need to be treated with antibiotics.

AVOIDING CONTACT: Like spiders and centipedes, scorpions are nocturnal creatures, hunting at night. They do not attack people and scorpion stings are very rare. If you are camping in the bush, do not leave you shoes outside overnight and shake out all clothing before putting it on.

WHERE TO WATCH OUT: Scorpions live under rocks, logs or any other objects lying on the ground. Some species prefer to live under the bark of standing or fallen trees. They are found throughout all regions of Australia.

Processionary caterpillar *Ochrogaster lunifer*

These caterpillars are sometimes called hairy caterpillars.

There are many species of moth that have hairy caterpillars but the processionary caterpillar is probably the best known. Hundreds of them march in a long line, touching each other while heading towards the base of a suitable tree where they *pupate*. Children find these processions fascinating and often stop the line to see whether the caterpillars will join up again or keep on marching in two lines. Touching the caterpillar with a bare finger is enough to pick up their hairs and end up with an itch.

SIZE: Each caterpillar measures about 5 centimetres long but the procession line may be many metres long.

SEVERITY OF INJURY: The hairs are not venomous but are sharp enough to penetrate the skin, even at the slightest touch. The bristles are also difficult to remove once they are stuck in the skin. Unaware of having the hairs on their fingers, people may rub their eyes or touch other parts of their body, moving the hairs and spreading the itch.

AVOIDING CONTACT: These caterpillars are not to be touched with bare skin or even with gloves. The hairs will stay in the gloves and can penetrate the owner's or someone else's skin later.

WHERE TO WATCH OUT: Processionary caterpillars are clearly visible when on the march but not so much when they form a bag of silk at the base of or up in a tree. As the caterpillars shed their skin and transform into *pupae*, the skins dry out, fall apart, and the hairs are carried away with the wind. If this happens, innocent bypassers can experience severe respiratory reactions from breathing in the hairs, and often do not realise where the problem came from.

Australian magpie *Gymnorhina gibicen*

The magpie is one of Australia's most familiar birds.

The magpie is a medium-sized bird with a solid, sharp beak. During the nesting season, magpies may think innocent walkers are enemies and attack them fiercely. Research has found that when both male and female birds are involved in the attacks, they regard people as nest robbers. If only the male attacks, he is showing the female how good he is at protecting their nest.

SIZE: Australian magpies grow to 38–45 centimetres long.

SEVERITY OF ATTACK: A magpie's aerial attacks are a nuisance more than a real danger. A magpie will swoop someone from behind, pecking their head with its sharp beak. Although these attacks rarely cause people to bleed, it is the sudden surprise that shocks people. Magpies do not only attack walkers but also unwary bike riders.

AVOIDING AN ATTACK: Magpies usually attack people walking in parks or along median strips. The easy way to avoid these pesky birds is by avoiding the areas where they nest. It also helps to wear a hat, bike helmet or some kind of a head cover, even a newspaper over your head will help! Magpies will not attack front on, so you could always try to trick them by painting a pair of eyes onto the back of your hat or wearing sunglasses on the back of your head!

WHERE TO WATCH OUT: Magpies are found throughout Australia. Be careful in springtime, when magpies nest — this is the only time of the year when they practise air raids!

harmful

harmful

Mosquito *Aedes aegypti*

The mosquito can cause many health problems.

Mosquitoes can carry a number of tropical diseases such as dengue fever, Ross River virus, Barmah Forest virus and malaria. Malaria was removed from Australia in 1981, but the dengue fever virus, which is carried by the *Aedes aegypti* mosquito, is a serious problem in northern parts of Australia.

SIZE: *Aedes aegypti* is a small mosquito with white markings and banded legs.

DISEASE SYMPTOMS: Typical health problems are fever, headache, pain in the joints and muscles, rash, nausea, and vomiting. Dengue haemorrhagic fever, dengue shock syndrome or meningoencephalitis can also occur.

AVOIDING INFECTION: The males do not bite people, only the female mosquitoes bite and suck blood. The mosquito is only the carrier of the disease. It must first bite and suck blood from an already-infected person and then bite another person within a short period of time.

To prevent such spread of disease, make sure that pot plant holders, and any other items in the garden or on the verandah that may contain water are regularly emptied, because that is where these mosquitoes breed. The use of personal insecticide and long-sleeved shirts are recommended when outside. Also, every house in the tropics should be fitted with insect screens.

WHERE TO WATCH OUT: Moist areas in tropical Australia make the best habitats for *Aedes aegypti* mosquitoes to live and breed. There can be millions of them during the wet season.

Spectacled flying-fox *Pteropus conspicillatus*

The spectacled flying-fox is one of Australia's megabats and a possible carrier of many diseases, including the deadly lyssavirus.

Flying-foxes may look cute but they can also carry deadly diseases. The lyssavirus, which bats may transmit to humans by biting, is related to rabies. All species of flying-foxes can carry lyssavirus.

SIZE: Head and body measures 22–24 centimetres.

DISEASE SYMPTOMS: Lyssavirus and rabies cause headache, fever, weakness and coma. Death follows several days later.

AVOIDING AN ATTACK: Flying-foxes often end up on the ground, particularly after heavy storms. They can also become tangled up in fencing wire. Whether injured or not, they should never be picked up with bare hands.

Call an experienced wildlife carer or the local wildlife authorities if ever you find an injured flying-fox. Even a dead flying-fox should not be handled. If it needs to be buried, don't touch it, scoop it up with a shovel.

WHERE TO WATCH OUT: The spectacled flying-fox is found in a limited range within the rainforests of north-eastern Queensland.

Feral pig *Sus scrofa*

Feral pigs are full of nasty parasites.

Feral pigs are not native to Australia. They come from domestic pigs that have escaped or were deliberately released. An injured or cornered pig may attack, causing deep, nasty cuts from the tusks. Pigs also spread diseases like tuberculosis. Feral pigs have been officially declared an "environmental pest".

SIZE: An adult pig can grow 1.5 metres long and weigh over 100 kilograms.

DISEASE SYMPTOMS: Internal parasites like tape, hook and round worms can cause serious damage to the stomach lining, liver and other organs, depending on the type of parasite. Tuberculosis is not common in humans but can be easily transferred to livestock, which could be disastrous for the cattle and domestic pig industry.

AVOIDING CONTACT: Feral pigs usually run away from people. However, an injured pig may defend itself. Not only can they cut and gore with their tusks, but if the pig's saliva gets into the wound, there is a possibility of passing on a disease or infection. It is very unsafe to eat meat from a feral pig, particularly the internal organs.

WHERE TO WATCH OUT: Feral pigs live throughout Australia, except for Tasmania. They prefer wet habitats, and when food is in short supply they invade crops and even suburban gardens. Feral pig meat is not sold in Australia but some pig hunters choose to eat what they hunt. If you are ever offered wild pig meat, politely refuse.

Ciguatera poisoning

Eating big fish, such as coral trout, can give you ciguatera poisoning.

Ciguatera poisoning is caused by eating large predatory fishes that live on coral reefs. The poison is generated in tiny plankton (dynoflagellates) that are eaten by small fishes, which are then eaten by larger fishes. The poison, called ciguatoxin, is stored in the fish's flesh, so fishes like barracuda, moray eel, mackerel, coral trout and others that have eaten many smaller fishes all the time may be full of ciguatoxin.

SIZE: The bigger the fish, the greater the chance of ciguatera poisoning.

DISEASE SYMPTOMS: The symptoms of ciguatera poisoning are peculiar. While there is no fever, the patient may feel generally sick, suffer from diarrhoea, have a metallic taste in the mouth, have sore calf muscles, lose their balance and sweat. These problems may last for several days and people who are unfortunate enough to eat fish with ciguatera again usually suffer even worse. Ciguatera is a poisoning, not an allergy.

AVOIDING AN ATTACK: It is not possible to know if a fish has ciguatoxin just by looking at it. There are now tests for checking if they have ciguatera, but these are done in a laboratory. To be on the safe side, buy fillets from smaller fish.

WHERE TO WATCH OUT: The Great Barrier Reef in Queensland is the main area for ciguatera, but other areas can also have poisoned fish. Ciguatera poisoning is the very reason why fish such as moray eel, barracuda and certain other predatory reef fish are not commercially harvested and are generally not eaten by people.

HOW MANY PEOPLE HAVE DIED?

It is not an easy task to tell exactly how many people have died in Australia from bites, stings and attacks caused by animals. We can guess that many people were killed by wildlife before European settlement and also that many casualties were not reported from remote communities in the past. Another reason for poor early records was the unreliable identification of animals because the classification system of Australian fauna had only just started.

Many fatal attacks were assigned to the wrong species or in some cases there were no witnesses to verify the details of the accident. Even in modern times, there is no requirement to keep records of deaths caused by wildlife but the Australian Bureau of Statistics does keep such records.

The figures presented in the table listing the deaths by various animals below, should be interpreted as "at least" statistics rather than absolutes.

	Common name	Scientific name	Australia	Worldwide
Mammals	Dingo	*Canis lupus dingo*	2	
	Water buffalo	*Bubalus bubalis*	1	
Birds	Southern cassowary	*Casuarius casuarius*	2	
Reptiles	Estuarine crocodile	*Crocodylus porosus*	17	
	Western taipan	*Oxyuranus scutellatus*	6	
	Common death adder	*Acanthophis antarcticus*	5	
	Tiger snake	*Notechis scutatus*	18	
	Chappell Island tiger snake	*Notechis scutatus serventyi*	2	
	Eastern brown snake	*Pseudonaja textilis*	19	
	King brown snake	*Pseudechis australis*	1	
	Rough-scaled snake	*Tropidechis carinatus*	1	
	Eastern small-eyed snake	*Cryptophis nigrescens*	1	
	Copperhead	*Austrelaps superbus*	1	
	Sea snakes	Various spp.	0	150
Amphibians			0	
Insects	Bee	*Hymenoptera* spp.	38	
	Wasp	*Hymenoptera* spp.	7	
	Bulldog ant	*Myrmecia* spp.	6	
Arachnids	Sydney funnel-web spider	*Atrax robustus*	13	
	Red-back spider	*Latrodectus hasselti*	14	
	Paralysis tick	*Ixodes holocyclus*	20	
Fish	White shark	*Carcharodon carcharias*	16	224
	Tiger shark	*Galeocerdo cuvier*	84	
	Bull shark	*Carcharhinus leucas*	2	71
	Hammerhead shark	*Sphyrna* spp.	0	16
	Bronze whaler shark	*Carcharhinus brachyurus*	2	12
	Stonefish	*Synanceia* spp.	1	4
	Stingray	Various spp.	3	17
Jellyfish	Box jelly	*Chironex fleckeri*	67	
	Irukandji	*Carukia barnesi*	2	
Other marine animals	Cone shell	*Conus* spp.	1	15
	Blue-ringed octopus	*Hapalochlaena maculosa*	2	1

Accidents involving wildlife can happen at home, in the backyard, on the way to school, while camping in the bush or swimming in the sea. Although we cannot always prevent accidents, we can at least lessen the risk of injury if we know what kind of dangerous animals we may see in certain places and situations. Knowing what those animals are capable of doing helps us prepare for any unpleasant encounters. Even the most dangerous animals will usually not attack humans — as long as they are not harassed or suddenly surprised by us.

Above and inset: Warning signs are there for a reason — take them seriously!

When working in the garden or collecting firewood, always wear gloves and look before you pick up any object that has been lying on the ground for some time. Snakes, spiders, centipedes and other critters like to hide under the cover of logs and leaf litter and can get quite annoyed when disturbed. Swimming, snorkelling or just playing in the water should not be spoilt by dangerous creatures lurking under the surface. In the tropics, where crocodiles and other dangerous predators live, never go into the water or even close to the edge if you cannot see the bottom. At some beaches, rivers and swamps, warning signs have been placed to let people know that box jellies or crocodiles live there. Please take such warnings seriously. Most popular beaches have lifesavers on patrol who will let people know of any potential danger. They will also have first aid kits and experience in helping people in trouble.

Always wear protective shoes when walking in the bush!

What about creatures such as disease-spreading mosquitoes? No matter how we try to avoid them, they are out there. To avoid most biting insects, we can protect ourselves with personal insect repellent, a long-sleeved shirt and long pants, mosquito nets and screens. However, it is best to avoid places that are thick with mosquitoes altogether.

When in the bush, wear closed shoes and, if possible, long pants. There is more danger on or near the ground, so keep your eyes open for snakes and other creepy crawlies. Outdoor activities are great fun, so do not spoil it for yourself with careless behaviour. We have to stay safe while sharing this planet with other creatures — even with the dangerous ones.

Make sure that you or someone else is not going to get hurt while applying first aid. Some snakes try to bite more than once, wasps can also sting repeatedly and being stung or bitten in the water could mean possible drowning. It is important to get away from the animal before providing first aid — this is especially important when helping another person. If the danger is still there, it could mean two victims instead of one.

Remember not to panic. Knowing how to apply the appropriate first aid is also important because the wrong first aid can cause more harm than good.

SNAKE BITE

Apply a pressure bandage and immobilise the limb (stop the affected limb from moving). Do not wash or wipe off any venom as this will help identify the type of snake. Do not kill the snake; however, if the snake is already dead, take it to the hospital for identification. Take the victim to hospital as quickly as possible.

SPIDER BITE

Because spiders have a different type of venom to snakes, different first aid is needed for spider bites. The best first aid method for spider bites is to place an ice pack over the bite and, if necessary, get the patient to a doctor.

However, a Sydney funnel-web spider bite should be treated in the same way as a snake bite by applying the pressure/immobilisation method, not the ice treatment! Quickly get the patient to the nearest hospital.

ANT, BEE AND WASP STINGS

Remove the insect. In the case of a bee sting, the sting with the venom sac must be quickly pulled out of the skin. Apply an ice pack or bathe the wound in icy water to help reduce the pain and swelling. If, however, the person is known to be allergic to these stings, the pressure/immobilisation method should be used to slow down the spread of the venom through the body. Professional medical treatment is absolutely essential in such a case.

SEA JELLIES

The sting from the deadly box jelly and the irukandji should be treated by pouring lots of vinegar over the stung area. However, if it is a Portuguese man-o-war (bluebottle) sting, place an ice pack or icy water on the area stung. It may also be necessary to take some strong pain killers. It is very important to get the patient to the hospital as soon as possible!

STONEFISH AND OTHER VENOMOUS FISH

It is very important to try and stop the terrible pain. It helps to put the foot or hand in warm water. The wound should never be cut or bandaged. Give the victim pain killers and get them to the hospital quickly.

PARALYSIS TICK

Remove the tick as soon as it has been discovered. Do not use any chemicals to remove the tick as it will irritate the tick and it will inject more venom. Use sharp-pointed scissors or fingernails. Avoid squeezing the tick! If the patient feels sick, medical treatment may be necessary.

CUTS, SCRATCHES, PUNCTURES

Injuries caused by non-venomous creatures such as sharks, crocodiles and other animals should be treated depending on how badly the victim is hurt. The patient should be made comfortable and kept as calm as possible.

Severe injuries may require pressure bandaging, others may just need cleaning and disinfecting the wound to prevent infection. In the case of an injured limb, it should be elevated to slow down the blood flow into the injured area.

GLOSSARY

Allergic Being very sensitive to certain things (such as dust or bee stings). Such people have an allergic reaction and can become very sick if they come into contact with these things.

Ambush A sudden and surprise attack.

Anticoagulants Chemicals that prevent the victim's blood from clotting.

Antivenom A serum used to treat people who have been bitten by a venomous snake.

Camouflaged To blend in with the surroundings using colours and patterns.

Cardiopulmonary resuscitation (CPR) By using mouth-to-mouth resuscitation and pressing down on the chest CPR is used to save lives in an emergency.

Carnivore An animal that eats meat and other animals.

Colony A group of the same kind of animals living together.

Disembowel To tear out the bowels or intestines.

Diurnal Active during the day.

Domesticated Animals that are tame and live with humans.

Dorsal Of the back (usually spines or fins on the back of a fish).

Extinct Having no living members of a species.

Fang A hollow or grooved tooth used to inject venom.

Fronds Large leaves like those of ferns and palms.

Habitat The specific place where a plant or animal lives.

Hibernate A state of inaction or sleep during winter or cold weather.

Keeled scales A raised ridge running lengthwise along the middle of the scales of a reptile.

Lethal Causing death.

Mandibles An insect's/animal's jaws.

Nematocysts Extremely small organs inside the cell of a sea jelly that cause a sting.

Neurotoxins Chemicals that attack the central nervous system. The victim is quickly paralysed and then dies if untreated.

Nocturnal Active during the night.

Paralysed Loss of control over the muscles.

Pectoral fins The side fins of fishes and sharks.

Pedicellariae The parts of a sea urchin that look like flowers.

Pedipalps Appendages on either side of the mouth of a spider.

Pelvic spines Spines on the underside of fishes.

Poison A substance that, when eaten or absorbed by a living organism, destroys life or injures health.

Predators Animals that hunt and eat other animals.

Primates Any mammals of the order Primates, including humans, monkeys, chimpanzees etc.

Proboscis Any snout-like feeding organ.

Procoagulants Chemicals that cause a victim's blood to clot.

Pupa (Plural: Pupae) Inactive stage between larva (caterpillar) and adult. Also known as a chrysalis.

Pupate To become a pupa.

Spur The tiny remains of legs that are found on some snakes (such as pythons).

Species A group of animals that share the same features and can breed together to produce fertile offspring.

Temperate Of medium temperature.

Venom A toxic substance made by animals that can be injected by fangs, spines or stingers into prey.

INDEX